A LOOK AT EARTH SCIENCE

OCEANOGRAPHY

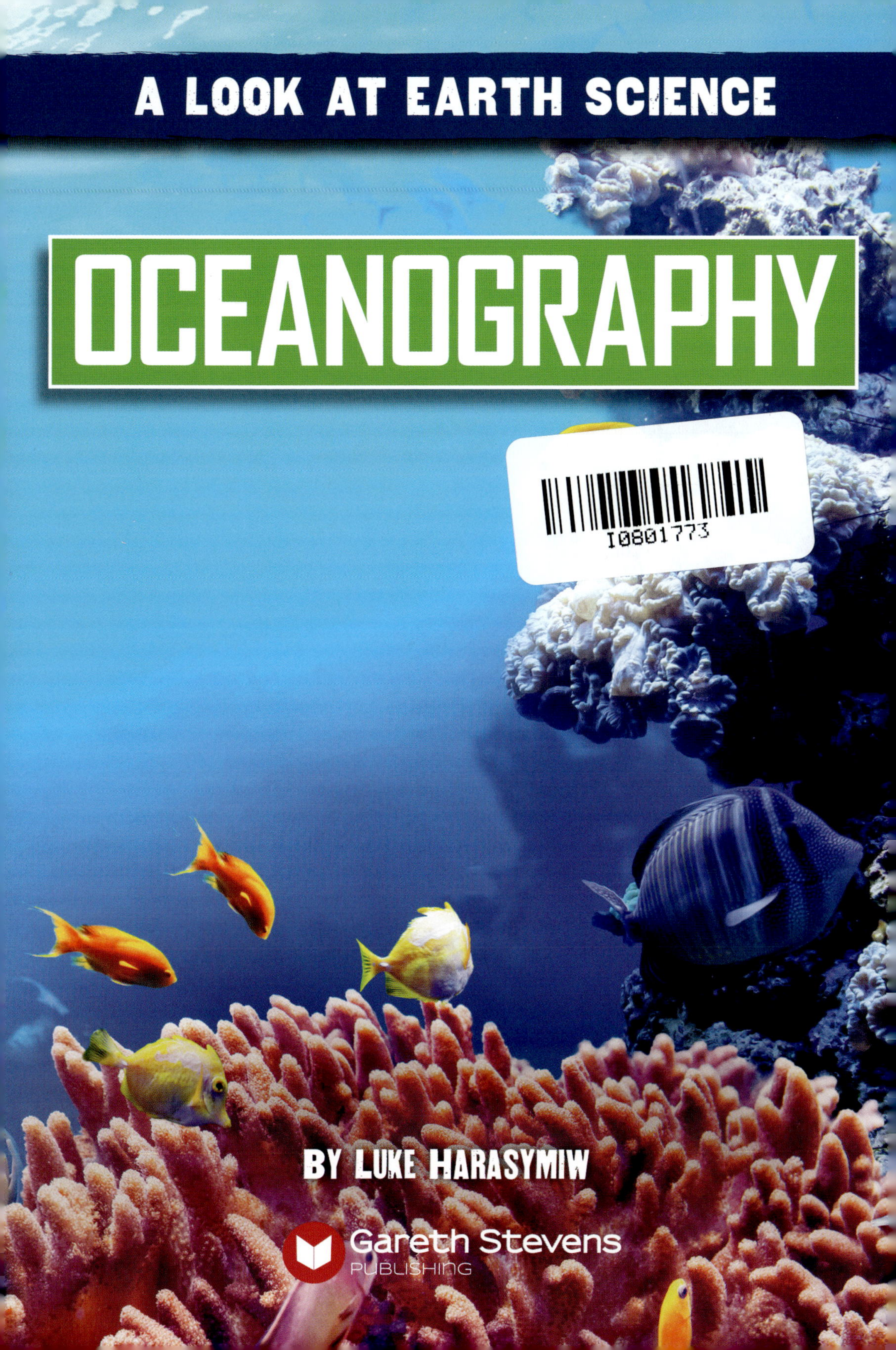

BY LUKE HARASYMIW

Gareth Stevens PUBLISHING

Please visit our website, www.garethstevens.com. For a free color catalog of all our high-quality books, call toll free 1-800-542-2595 or fax 1-877-542-2596.

Cataloging-in-Publication Data

Names: Harasymiw, Luke.
Title: Oceanography / Luke Harasymiw.
Description: Buffalo, NY : Gareth Stevens Publishing, 2025. | Series: A look at Earth science | Includes glossary and index.
Identifiers: ISBN 9781482467260 (pbk.) | ISBN 9781482467277 (library bound) | ISBN 9781482467284 (ebook)
Subjects: LCSH: Oceanography–Juvenile literature. | Ocean–Juvenile literature.
Classification: LCC GC21.5 H375 2025 | DDC 551.46–dc23

First Edition

Published in 2025 by
Gareth Stevens Publishing
2544 Clinton Street
Buffalo, NY 14224

Designer: Jennifer Schoembs
Editor: Therese Shea

Photo credits: Cover silvae/Shutterstock.com; series art (backgrounds) Andrey_Kuzmin/Shutterstock.com; p. 5 loskutnikov/Shutterstock.com; p. 7 (map) Pyty/Shutterstock.com; p. 7 (bottom) Kelly Headrick/Shutterstock.com; p. 9 (map) WindVector/Shutterstock.com; p. 9 (bottom) BogDarena/Shutterstock.com; p. 11 Nicole Helgason/Shutterstock.com; p. 13 BlueRingMedia/Shutterstock.com; p. 15 Jose Angel Astor Rocha/Shutterstock.com; p. 17 Sascha Caballero/Shutterstock.com; p. 19 NOAA/Wikimedia Commons; p. 21 Vlad61/Shutterstock.com; p. 23 Bizi88/Shutterstock.com; p. 25 Petty Officer 1st Class Peter Lewis/Wikimedia Commons; p. 27 Lukasz Z/Shutterstock.com; p. 29 divedog/Shutterstock.com.

Printed in the United States of America

CONTENTS

Words in the glossary appear in **bold** type the first time they are used in the text.

OCEANS AND MUCH MORE

Oceanography is the study of Earth's oceans. That includes ocean water, ocean movements, the effects of the oceans on the planet, and the life-forms within the oceans. It's a really big area of science! The oceans are really big too. About 70 percent of Earth's **surface** is covered by oceans.

MAKE THE GRADE

Oceanography is a branch of Earth science. Earth science is the study of our planet and its atmosphere, or the gases that surround it.

SO, WHAT'S AN OCEAN?

An ocean is a huge body of salt water. Earth has five oceans: the Atlantic, the Pacific, the Indian, the Arctic, and the Southern, sometimes called the Antarctic. These marine, or ocean, waters don't have borders. So really, there's just one world ocean!

MAKE THE GRADE

How big are the world's oceans? They contain about 98 percent of Earth's total water!

Oceanography also includes the study of smaller bodies of salt water that connect to the oceans. These include seas, **gulfs**, bays, and **straits**. Seas are usually partly surrounded by land. The Sargasso Sea in the Atlantic Ocean has no land around it, though!

MAKE THE GRADE

Scientists call the Sargasso a sea because the ocean currents, or large flows of water, within it separate it from the Atlantic Ocean. It's named for the seaweed called sargassum, which floats in it.

PHYSICAL OCEANOGRAPHY

People who study oceanography are called oceanographers. The study is so wide that oceanographers often **focus** on one part. Physical oceanography is the study of ocean water. This includes its temperature, or how hot or cold the water is, in different places.

MAKE THE GRADE

The farther down you go in the ocean, the more **pressure** increases, or rises. That makes it harder for oceanographers to study deep under the surface.

Some physical oceanographers focus on the water's movements, such as currents. Ocean currents move a bit like rivers within the ocean. They help control **climates** around the world. They move heat from the **equator** to the poles.

MAKE THE GRADE

Some ocean currents (shown as C. on this map) flow at the water's surface. Others flow deep under the water. Some are large rotating, or circling, currents formed by winds and other forces called gyres (JY-uhrz).

CHEMICAL OCEANOGRAPHY

Some oceanographers study **elements** found in ocean water. This study is called chemical oceanography. Water is made up of the elements oxygen and hydrogen. Ocean water also contains sodium chloride, or salt, which is made of the elements sodium and chlorine.

MAKE THE GRADE

The ocean's salt mostly comes from rocks and seafloor openings called vents.

Ocean waters also contain smaller amounts of other substances, or kinds of matter. Carbon dioxide is a gas absorbed, or taken in, by the oceans. A certain amount is helpful and used by marine life. But increasing amounts of carbon dioxide are harming marine life.

dying coral

MAKE THE GRADE

The human-caused increase of carbon dioxide in the atmosphere is raising carbon dioxide levels in the ocean. It's changing the chemistry of the water, an effect called ocean acidification, which can harm marine life.

UNDERWATER GEOLOGY

Marine geologists study ocean features, such as the floor, **continental shelves**, and hydrothermal vents. Hydrothermal vents are openings in the ocean floor from which hot water flows. Ocean water leaks into the earth, is heated by hot, liquid rock called magma, and comes out a vent.

MAKE THE GRADE

The water that comes out of hydrothermal vents can be 700°F (370°C)! Some ocean life-forms, such as yeti crabs, only live near these superhot openings.

OCEAN ECOLOGY

Ecology is the study of plants and animals. Marine ecologists study ocean plants and animals. There are more than 228,400 species, or kinds, of ocean animals. These include about 130 species of **mammals**, including whales and seals, and about 20,000 fish species!

MAKE THE GRADE

Scientists think it's possible that more than 90 percent of ocean species haven't been discovered yet!

Scientists who choose marine ecology may study one species or a certain area. They may study how pollution, or harmful matter put into the water, affects marine life. They may focus on an unexplored area. In 2024, scientists found four new species of octopus near Costa Rica!

MAKE THE GRADE

Another name for marine ecology is biological oceanography.

WORKING TOGETHER

Some oceanographers take samples of water or ocean life and study them in labs. Others work mostly on and in the oceans. They spend a lot of time on boats. They may even take underwater vehicles deep under the surface.

MAKE THE GRADE

Some oceanographers use underwater robots to explore places too dangerous for people to go!

OUR AMAZING OCEANS

The oceans provide people with so many things. People use oceans to travel from one place to another. They need them for food and for jobs. Some need them for water too. They may use a process, or method, called desalination to take the salt out of the water.

oil platform

MAKE THE GRADE

People get **minerals** they need from the ocean floor. They also get oil and gas from beneath the continental shelf.

Oceans help people in so many more ways. Substances found in marine life, such as coral, help fight **cancer**. Certain marine bacteria may help clean oil spills in the water. More marine discoveries happen each year, thanks to oceanographers!

MAKE THE GRADE

According to scientists, about 95 percent of the ocean is still unexplored.

RECENT OCEANOGRAPHY BREAKTHROUGHS

In 2024, a seamount, or underwater mountain, two times as tall as the world's tallest building was found near Guatemala.

Scientists finished a map in 2023 that shows more than 19,000 underwater volcanoes.

A study completed in 2024 showed fish grow larger when they're away from places where fishing is allowed.

A remote-controlled submarine found the tooth of a megalodon, an ancient giant shark, in the Pacific Ocean in 2023.

Underwater worms called tubeworms discovered near Alaska in 2023 may live to be 200 years old.

Scientists found tunnels beneath the ocean floor, and marine creatures living in them, in 2023.

A satellite was sent into Earth's atmosphere in 2024 to study the oceans from above.

GLOSSARY

cancer: A disease caused by the uncontrolled growth of cells in the body.

climate: The average weather conditions of a place over a period of time.

continental shelf: The part of a landmass called a continent that lies under the ocean and slopes to the ocean floor.

element: Matter that is pure and has no other type of matter in it.

equator: An imaginary line around Earth the same distance from the North and South Poles.

focus: To direct attention or effort to something specific.

gulf: A large ocean area partly surrounded by land.

mammal: A warm-blooded animal that has a backbone and hair, breathes air, and feeds milk to its young.

mineral: Matter in the ground that forms rocks.

pressure: A force that pushes on something else.

satellite: An object that circles Earth to collect and send information.

strait: A narrow channel of water that connects two large water bodies.

surface: The top layer of a planet.

FOR MORE INFORMATION

Books

Cusolito, Michelle. *Diving Deep: Using Machines to Explore the Ocean.* Watertown, MA: Charlesbridge, 2022.

Strauss, Rochelle. *The Global Ocean.* Toronto, ON, Canada: Kids Can Press, 2022.

Website

National Ocean Service: For Kids
oceanservice.noaa.gov/kids/
This U.S. government agency has many links about oceans and ocean life.

Publisher's note to educators and parents: Our editors have carefully reviewed these websites to ensure that they are suitable for students. Many websites change frequently, however, and we cannot guarantee that a site's future contents will continue to meet our high standards of quality and educational value. Be advised that students should be closely supervised whenever they access the internet.

INDEX